NOTES FOR CONTINUING THE PERFORMANCE

Notes for Continuing the Performance

~ Anne Pitkin ~

Grateful acknowledgment is made to the editors of the following periodicals where some of these poems first appeared:

The Nation, Antigonish Review, Malahat Review, Prism International, Mill Mountain Review, Kansas Quarterly, Gilt Edge, Jawbone, The Niagara Magazine, Poetry Northwest

Copyright © 1977 by Anne Pitkin

ISBN 0-918116-04-X

JAWBONE PRESS
17023 5th Avenue Northeast
Seattle, Washington 98155

for Nelson Bentley

MAPS

for William Stafford

I know this country
as I go along, and leave
my footprints to what startled
ant, or mouse, or hawk
will scratch them out.

A seed pod, final as old age
is easily cracked. I'll dig
its kernel out and plant it,
hoping to discover, in another
season, somewhere else, what messages
begin with such a gesture.

I learn rain
by its collisions
with the lake, each flying
chip of glass a greeting
urgent as a rush of birds
toward a newly seeded field,
an invitation.

Loud as rain, where no one hears,
the moon drips, leaf by silver leaf,
into the lake, until its dazzled currents
swarm beyond their banks to settle
in a labyrinth of shadow

where I lose them,
where it's just a little darker
than it was the last time, as I learn
again to find the charts in their revision,
in each new relinquishment toward a perfect
starting point.

GRAVITY

The jet streaks through the morning
haze in slow motion, circles
in a wide arc like an arrow's,
while we gather at the terminal
and wait. It seems to graze
the mountain, disappears, returns
and circles.

In the heart of town an evangelist
falls silent on his corner. Sparrows
chattering around his head
have stopped mid-flight. He waits
for them to fall, prepares
an explanation, but they vanish.
Time to pack up

visions of what may descend
and bring us news. Time
for rained on leaves
that trembled in a brief fury
of sunlight to burn black
and silent in the evening air.
Soon afternoon will fall
as suddenly as blood falls
from a startled face. Listen

for the thunder
like approaching hoofbeats.
You won't hear it. One more time

the sun has left its light behind
in streaks that tear across the hills
and disappear without a sound.
A small plane like a silver fish alone
in all its element leaps up and spirals

distant, bright, incongruous
as hope, its light the only movement
in a sky not dark enough
for stars, its light the star
that falls and does not break
but circles closer, flashing
signals we don't understand.

DECEMBER

The sparrows flying off the page
appear in the background, because
the tree was erased, and the casual design
of thin clouds couldn't be exactly reproduced.
Now we agree, reluctantly, the sparrows
are better. We learn these failures early

struggling with the music, until
we have it mastered and we find we've lost
that first keen tenderness that drove us on.
Still, the same tones from the belltower vibrate
through all changes in the weather, traveling
at perfect intervals, although they toll

grief or joy, depending on the hour we hear them.
This last day of the month is cold and clear,
an open book's unbroken spine. The trees
along the boulevard repeat themselves and tangle
delicate as nerves in a forgiving sky.
The bare frame of a new house turns ruddy

in the sun's evening slant. Inside
she meets her lover for the last time
takes his arm and walks him slowly through
the rooms she and her husband have long planned.
A galaxy of dust defines each beam
of bronze light falling, as she passes,

on her radiant hair. Outdoors, on one patch
of sky, a small constellation of birds like smoke
loses shape. Last night's fresh snow quiets the street,
except for the staccato barking of a dog,
the distant shouts, like sparks from a struck match
of children rushing through the perfect air.

FAILED APHORISM FOR A SMALL SON

for Paul

It was out of control
cold, brilliant with anger.

Today you fling out through the kitchen door
toting a machine gun made of tinker toys.
This time you mean it. You get past
the mailbox, the vacant lot next door —
you hesitate on the corner . . .

New Year's Eve we left together
over the tight snow, holding hands and dragging
the sled across a fierce new planet. Lost
in a white storm frozen at its peak,
we climbed carefully along the edge
of the neighborhood. At the top I told you

Look back there and see how pretty, how it falls
away, the broken crystal glinting
in the folds of a tablecloth, and sparks blooming
from a thousand chimneys, festivals
of last words. We jumped on the sled. We got past
the crowds of shadows whistling in our ears.
I held you with arms brittle
as an eggshell while we tumbled, whooping,
front, sideways, and backward, our dark
elements wind and thin ice.

I know this much about love.

You'll come back in a minute,
still snuffling, to point the gun
exactly between my eyes. You'll watch, baffled
as it crumbles, piece by piece, until you're roaring
with chagrin, you, trying not to laugh, you,
tightening your grip on what's left, some odd
sculpture we can't put a name to, never
would have made on purpose.

BABY GORILLA'S ACCIDENT, A STORY FOR CHILDREN

Kongo (Daddy) is one cage.
Lulu (Mommy) is in the other cage.
Honeycake (Baby) is in Lulu's cage.
Honeycake reaches into Kongo's cage
and takes his hand. By and by

Lulu says I am going to take
Honeycake to the other side
of my cage. Kongo does not hear.
Kongo continues to hold Honeycake's hand.
Lulu does not see.
Lulu takes Honeycake's other hand.
Mommy pulls. Daddy pulls.

This is how the accident happens.

Baby splits down the middle.

Daddy is upset.
Mommy is upset. Mommy says why

don't you ever listen? Daddy says
I am sick and tired
of the configuration of this zoo.
I am sick and tired
of being locked in
these cages side by side
I am sick and tired
of your style why don't you change
your style?

Mommy says that's not what we were talking about.

Daddy says can you or can you not change
your style yes or no? Baby gets bored

picks herself up and puts herself together again just a little askew. Mommy and Daddy make an adjustment here, a compromise there, and congratulate themselves on a job reasonably well done.

WAITING IN THE CAR

Only the street lamp like a perfect moon
holds steady in this struggle of the leaves
against a sudden wind. If you slam the door
harder than usual I don't ask why

or watch you hurry through this frozen light
that breaks against the shadows
swinging overhead. A maddened swallow
pitches through the darkness, wrenches free

then lays its wings like scalpels
on the fading resonance of afternoon.
A man and woman argue soundlessly
behind a window. In that lighted frame

they seem to see each other. Hands fly up
between them, drop, then reappear in random
necessary gestures. A child's voice laughs
or cries. The swallow's flight grows

small against its changing context.
This is the street we live on. You and I
have yet to meet face to face
in the same angle of light, traveling

as we do under this shrinking moon
that is pulled like a bitten fingernail
across the limbs of all these trees
ruptured again and again into spring.

CHARTING OLD TERRITORY

Lines are drawn between us
on this map, resonant as strings
that quiver on a plucked
violin, taut as latitudes
the last crow in December draws
across fields long since cut.

On this empty pavement, luminous
with rain, a yellow leaf lies
like an open hand. Longing mends
the rupture between leaf and twig
limb and trunk, imagines roots
in friendly ground where forests spring,
wild with beasts inviting us to name them.
There you are

in amber light as silent
as the clear eye of a cat
poised for an instant in a pool
of sunlight. She never moves.

The stars around her leap
and plunge, are locked in flight
as certainly as those dark birds that pull
a corner of the clear sky south
from here, as certainly as you and I

are governed by the dark
between worlds long ago burnt out
whose light still travels, stubborn
courier of promises
not made, not broken.

ADDICTION

It makes no demands at first,
that quick play of sunlight
on the parked cars down the hill,
no demands, the yellow swarm
like bright hair tumbling

from the birches. Now it spreads
along the ground as if to keep the darkness
down forever. Creeping
into the hollows, it collects
in pools under doors locked for years

on your empty rooms. And the house
you live in, your dim house
comes alive with the fragrance of dead flowers
overflowing their vases. Time
to replace them, start new plants

on the sill. In these warm days,
these large rooms building themselves out
into the winter, roots spread
and knot, and demand more room.
The decorous shoots climb, weaving

themselves hour by hour into a sweet
tangle of limbs in the once discreet
windows. They shield you from the glare
of daylight. You are free
to absorb yourself in the walls

and their intricate shadows. Under the eaves
webs are spinning toward perfection by this time
of year, and the torn fish battle
upstream against the river running itself
out and out along its ravenous course.

GLASS HOUSE

You sleep. I watch the altered sky.
Those branches have been pruned
where once meteors like fireflies
tumbled without breaking, limb to limb
through darkness that was kind

as strangers. We are maimed by need
relentlessly fulfilled. You reach for me,
wake up, and pull away. Stars fall slowly
down the window, exquisite as snow
against a warm coat. What has been

nourished and contained now paces,
hungry, sleepless, in the separate rhythms
of our breath. All day long I've watched
white summer moths like wind rattled
blossoms close and open in the evergreens

then drift upward, unchanged, never
stripped, or hardened into fruit.
Love holds us here, love, that rock
we throw repeatedly to break the silence,
hope that scatters stars like startled

minnows out of old formations. We wake up
each morning, short of dawn, to cadences
of footsteps that recede, to rain
that crumbles down around this house
where we are never quiet, seldom kind.

E PLURIBUS UNUM

His hands do not know
that he's a fine musician.

His feet know that he's
a dancer, and his legs

do not.   So much for that
you say.   My hands don't know

right from wrong, so I had one
removed, but only decreased

my crime rate by 50%.   I say
that's *your* problem.   We the faithful

kneel on the stone floor
of the chapel.   We have not

had breakfast.   The soul knows
where it wants to go, is not

distracted.   A chair falls
over.   Our ribs joggle.   The soul

is not distracted, rises
with the heat.   A sparrow rises

from an organ pipe, rides
a glissando up, up, and bats

at the rafters.   Our heads begin
to sing.   This is what we've

been waiting for.   A moan
uncoils beside the fallen

chair. The body does not know
the soul is steadfast. The body

faints. Outside the wind knows
where it's going, argues about it.

THE FOOTBRIDGE

like an empty clothesline is slung high
above the widest section of the river
between two shores that look alike. I curse
at having found it and start across.
I have a choice between gripping a string
of cable at ear level or one at hip level.
I don't look down. I place one foot exactly
in front of the other, looking only ahead
to the moment when I can distinguish a face
on the other side. I seem always to drift up
stream sideways. I never meant to come
this way — one foot primly in front of the other —

but I came because I was afraid
not to. I inch forward
because I am afraid
to look back. I am afraid
someone I've left behind
will jump on the end of this toy
this absurdly jointed snake, causing
it to gallop, to throw me — and the river
the shallow blue river swarming
with sunlight dashed to bits
but not diminished

has long since rejected me.
I don't think of ruptures, knowing
no amount of preparation will
minimize the snapping
the tearing . . . now it comes
it comes again

the wind, swearing, drunken
fisherman, dropping nets all over

the water, lines tangled,
flailing at obstacles,
and there is nothing
for it but to hang on
hang on, knowing this
is what I came for.

REPATRIATION, A BEGINNING

You have fallen out, a vine
tumbling like too much joy
over the broken frame, a lost
precision of sunlight on the floor.

Having loosened the sky into flight
the breath of your life
lays bare the limbs
of trees, and the nests protected
like small hearts tremble
apart twig by twig.

The lights are going on
in all the houses, and children
are gathered in for the evening.
The ground you stand on quakes
a little as they run. There is nothing

for you to learn just now
except how minutes grow
upon each other, flowering simply
as a stray beam flowers
through a stained glass window.

Your coat has dropped
and the keen passage of the seasons
probes the loose threads
of your old defenses.
In a house that is not yours

the original statement of a fugue
grows past itself
like the tributaries of a river
the lines on a face.
You are not the musician.

The lights in the valley reach into the dark
from their symmetrical clusters.
They are far from you.
But you can count on them
and on the stars that rush to fill
the rising emptiness — that emptiness
the heart laid open to its next beat.

SOME NOTES ON THE HOLY GRAIL

The search for the Holy Grail has become
In. This represents a clear advantage
to the nation, because it keeps a lot of people busy
who would otherwise glut the market
on wealth, fame, sex, honor, war, art
and all the rest of it. These are only
intermittently tangible. The Holy Grail

on the other hand, is an eminently
worthwhile thing. It is something
you can hang on to. When you get bored
with it, you can go after another one.
If you really hustle, you can accumulate
a large collection, which you can then
polish, arrange, and rearrange
like a tea set. When you feel you need

to expand your need to expand
your unexpected potential, you can enter your
                                                collection
in Holy Grail competitions.
You can go on Holy Grail cruises
in the Caribbean, the Mediterranean
or the Dead Sea, if you like.

You can write a book and talk about it
on television, thereby making enough money
to finance an extensive research project
into the origins of the Holy Grail,
for which attempt you may
become famous and be surrounded
by beautiful women and/or men

or you may not. What is important, of course
is the spirit in which you proceed
on which depends the degree to which

the Holy Grail will entertain you in your old age.

POINT OF REFERENCE

Low tide. The sun
brings the day down
to our level, then slowly
comes unraveled. Facing out
to sea he tilts his chair back,
a lone hieroglyph
against the growing margin.

We are dazed by the shine
of daylight at our feet.
Beyond us lies a country luminous
and veiled as possibilities that smoulder
in the dark and airless closet
of a seed. We head out, certain
that the glassy sand will hold us.

We come to our senses.

He has not moved. We fix our sights
on him for balance, steady
as he is on two thin legs
that hold in place the blurry shadow
he precisely rests on.

THE PIANO PLAYER ON THE 10th FLOOR

He is angry at the grand piano
he can't play for the thin walls
of his apartment high above the city.
His hands, those tongues ripped from their bells,
yank the curtains shut against the fog
that sinks into the street, a storm

that's lost its wind. A storm
is squirted on the ornamental piano
he keeps polishing. His glasses fog.
He bumps his head against the wall.
Upstairs someone drops a bar bell.
No one is safe in this city.

His elbow hits the keyboard. A city
thunders in his head. The storm
breaks in his hands. Dinnerbells
rattle their tongues. The piano
raises its voice. On every wall
a picture jerks askew. Someone's fog

cutter spills on the carpet, shattering the fog
a dinner party's settled into. A city
father, counting peacocks on the new wall
paper, drops his bifocals, and a storm
whirls, wringing its hands, toward the piano
heard but not seen. Oh you tenants ring the
                                        doorbells!

ring them loud and festive as the bells
of New Years tolling out the old fog
tolling in the new. He hears only his piano,
this beloved exile readmitted to his city.
Now all citizens arise and storm
the corridors. They pound the walls

until the walls

of the entire building crack like bells
of windchimes in a storm
that blooms as rage blooms, finally to crumble in a fog
that opens like a vast umbrella on the city.
Still, he plays his piano.

A storm is huddled out beyond the fog
that walls us in our city,
deaf to bells and someone's mute piano.

NOTES FOR CONTINUING THE PERFORMANCE

The brown wings twitch in the grass.
This is enough for now. Enough
to know the marrow of this battered limb
is green. Listen. They are playing

the concerto. They perform the music
as they give their lives to it,
although the concert hall is empty.
Where sunlight once fell simply

in the window, leaves like hands clapping
in another room rattle against each other,
can't let go of winter branches
creeping inch by inch across the glass.

A dead moth flutters in the screen.
A stunned bird doubts the air
will ever hold again. Still, we know
the continuity that agitates the hands

that stirs the useless wings, has kept
the small heart ticking in its rigid case.
It has happened many times before,
the shedding of the tough leaves

when invisible buds like fists unclosed
and pried them off. More than once
we've waked up, given nothing more than habit
to that green, that brief applause.

JAWBONE POETRY CHAPBOOK SERIES

#1   Sean Bentley

Out of the Bright Oasis: The Great Knight Reason

#2   Lois Lindblad

Learning to Swim

#3   David Brewster

The Child at the Bottom of the Lake

#4   Anne Pitkin

Notes for Continuing the Performance

Anne Pitkin was raised in Tennessee, and received both her B.A. and M.A. Degrees from Vanderbilt University. She moved to the Pacific Northwest in 1965, and now lives in Bellevue, Washington, with her husband and three children. Although Anne has only been writing poetry for six years, she's managed to achieve an impressive publication record, and has won several significant prizes, including the 1975 Pacific Northwest Writers Conference Poetry Award. This is her first published collection.

250 copies have been photo-offset and sewn by hand into paper wrappers. Type was set by Charlotte Casey. Chapbook design by Samuel Green and Sara Birtch. The cover illustration is from a pen & ink by Debbi Lunz.

25 copies, numbered and signed by the poet, are available at $4.00 each.